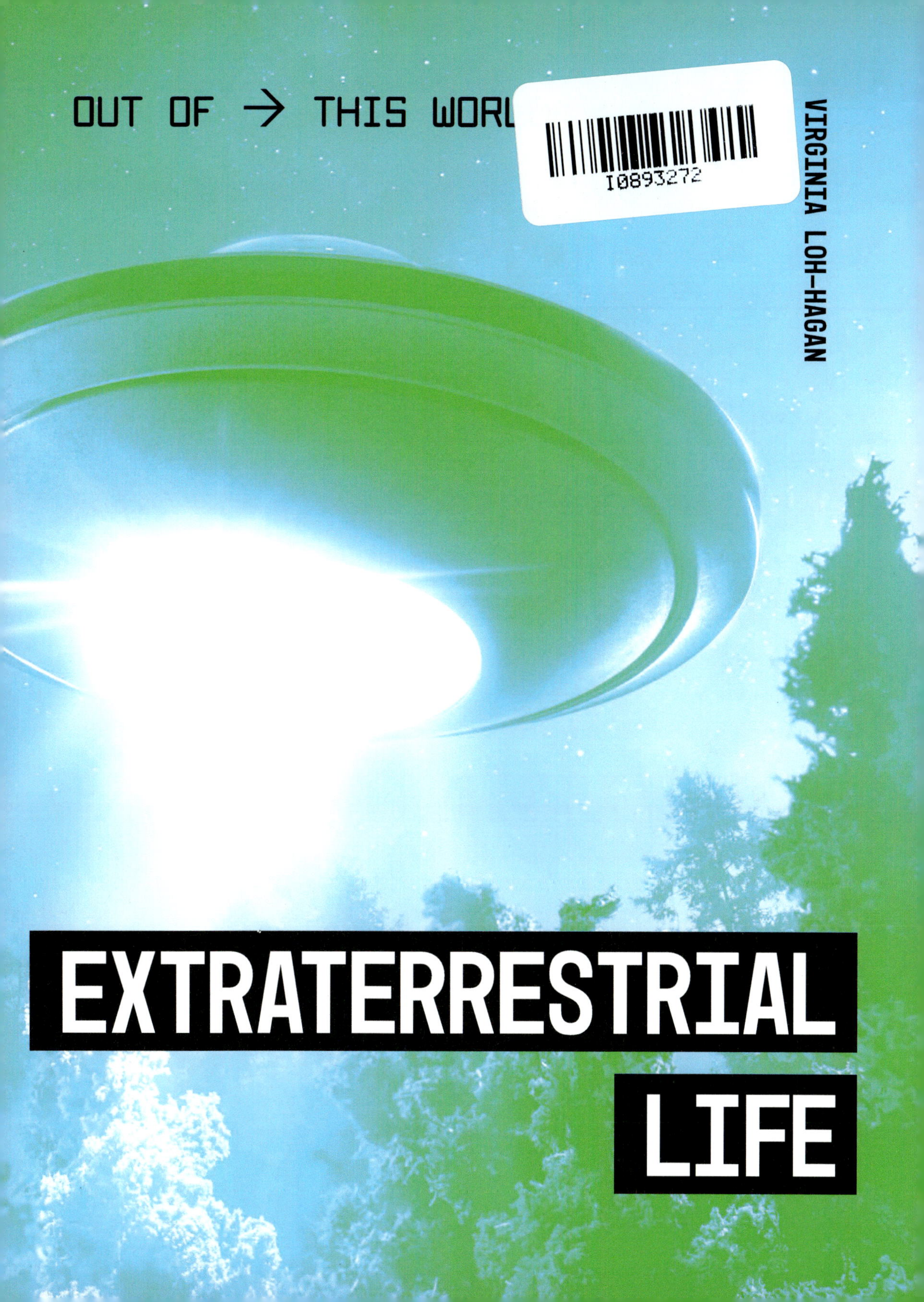

OUT OF → THIS WORLD

VIRGINIA LOH-HAGAN

EXTRATERRESTRIAL LIFE

45TH PARALLEL PRESS

Published in the United States of America
by Cherry Lake Publishing
Ann Arbor, Michigan
www.cherrylakepublishing.com

Reading Adviser: Marla Conn, MS, Ed.,
Literacy specialist, Read-Ability, Inc.
Book Designer: Jessica Rogner

Photo Credits: © Color4260/Shutterstock, cover, 1; © Vadim Sadovski/Shutterstock, 5; © NASA Goddard/NASA ID: PIA11980, 7; © NASA/GSFC/Chris Gunn/Flickr, 8; © NASA/Rick Guidice/NASA ID: ARC-1977-AC77-0475-3, 11; © NASA Goddard/NASA ID: PIA11226, 12; © NASA Goddard/NASA ID: PIA17659, 15; © NASA Goddard/NASA ID: PIA01486, 16; © NASA/JPL-Caltech/NASA ID: PIA22081, 19; © liquid studios/Shutterstock, 21; © IgorZh/Shutterstock, 23; © Tajborg/Shutterstock, 25; © Raphael Perrino/Flickr, 27; © NASA/NASA ID: PIA04216, 29
Graphic Element Credits: © Trigubova Irina/Shutterstock

45th Parallel Press is an imprint of
Cherry Lake Publishing Group

Library of Congress Cataloging-in-Publication Data
Names: Loh-Hagan, Virginia, author. | Loh-Hagan, Virginia. Out of this world.
Title: Extraterrestrial life / by Virginia Loh-Hagan.
Description: Ann Arbor, Michigan : Cherry Lake Publishing, 2020 | Series: Out of this world | Includes bibliographical references and index.
Identifiers: LCCN 2020006895 (print) | LCCN 2020006896 (ebook) | ISBN 9781534169289 (hardcover) | ISBN 9781534170964 (paperback) | ISBN 9781534172807 (pdf) | ISBN 9781534174641 (ebook)
Subjects: LCSH: Life on other planets—Juvenile literature.
Classification: LCC QB54 .L65 2020 (print) | LCC QB54 (ebook) | DDC 576.8/39—dc23
LC record available at https://lccn.loc.gov/2020006895
LC ebook record available at https://lccn.loc.gov/2020006896

Printed in the United States of America | Corporate Graphics

TABLE OF CONTENTS

4 . . . . INTRODUCTION
What Is the Universe?

6 . . . . CHAPTER ONE
What Is Extraterrestrial Life?

10 . . . . CHAPTER TWO
Is There Life on Other Planets?

14 . . . . CHAPTER THREE
Is There Life on Moons?

18 . . . . CHAPTER FOUR
What Is Tabby's Star?

22 . . . . CHAPTER FIVE
Have Aliens Made Contact?

26 . . . . CHAPTER SIX
Who Promoted the Study of Extraterrestrial Life?

30 . . . . GLOSSARY

31 . . . . . . FAR-OUT FACTS / LEARN MORE

32 . . . . . . INDEX / ABOUT THE AUTHOR

WHAT IS THE UNIVERSE?

The universe is huge. It's everything that exists. This includes planets, stars, and outer space. It includes living things on Earth.

The universe contains billions of galaxies. Galaxies are huge space collections. Galaxies are made up of billions of stars, gas, and dust. Galaxies include **solar** systems. Solar means sun. Earth is in the Milky Way galaxy. Galaxies spin in space. They spin very fast. There's a lot of space between stars and galaxies. This space is filled with dust, light, heat, and rays.

Before the birth of the universe, there was no time, space, or matter. Anything that takes up space is matter. Matter can exist in different states. The common states include solid, liquid, and gas. This is why things like air and smoke are considered matter. But the heat and light from a fire aren't matter. These don't take up space.

The universe hasn't always been the same size. It also hasn't always existed. Some scientists believe it began with a "big bang."

Galaxies travel through space. The Milky Way travels around 343 miles (552 kilometers) per second!

This is a **theory**. Theory means an idea. This theory explains how the universe was born. First, the universe was a super tiny blob, smaller than a pinhead! Then, that super tiny blob exploded. This happened 13.8 billion years ago. Next, energy spread out. Energy is made from matter. For example, the flames in a fire are matter. They take up space. But the heat you feel and the light you see from the flames are energy. Last, stars and planets formed. This all happened in less than a second.

Scientists think the universe is still expanding. Expanding means growing or spreading out. Scientists also think this expanding process is speeding up.

CHAPTER → ONE

WHAT IS EXTRATERRESTRIAL LIFE?

Humans have been searching for **extraterrestrial** life for years. Extraterrestrial means not from Earth. Extraterrestrial life could be space aliens. It could be **microbes**. Microbes are tiny organisms. Earth microbes can survive in extreme places. This makes people believe in extraterrestrial life.

The universe is really, really big. Some people think life has to exist beyond Earth. They think the big bang spread life to other places. Some think there are **habitable** zones around stars. Habitable means able to support life. These zones must have water. They must have breathable air.

Extraterrestrial life is also called intelligent life. Intelligent means smart.

Astrobiology is the study of life in the universe. These scientists study environments that support life. They study space science. They use many tools. They use space shuttles. They use **rovers**. Rovers are robots that explore the surface of space objects. Scientists use radios. They listen for signals. They use telescopes. Telescopes are tools used to see faraway things.

So far, scientists have only been able to study the Milky Way. They haven't found proof of life outside of Earth. They haven't been able to reach galaxies beyond ours. But they've found proof of **exoplanets**. Exoplanets are planets orbiting stars outside of the Milky Way.

Astrobiologists study life on Earth and in space.

AMONG THE STARS: WOMEN IN SCIENCE

Dr. Penny Boston graduated from the University of Colorado Boulder. She works at NASA's Astrobiology Institute. NASA stands for National Aeronautics and Space Administration. Aeronautics is the science that deals with the building and flying of aircraft. Dr. Boston serves as the director. She leads projects to search for life on other worlds. Her expertise is in caves. She studies caves on Earth. She also studies caves on Mars. She developed the idea of small jumping robots for exploring Mars. She gave a TED Talk about life on Mars. She studies organisms that live in extreme environments. She also writes poems. She said, "When it comes to searching for fellow creatures in the cosmos, we are armed with two primary tools. First, what we understand about the planet that gave us life. And second is one of our greatest inventions, the powerful tools of deductive and inductive science. Together, these comprise our handbook for understanding nature. The only tangible scientific model that we have to help us look for life on other planets is our biosphere here on Earth." Deductive means to make a general assumption about something and then prove or disprove it by observing specific things. Inductive means to observe specific things and then make general assumptions. Biosphere is the part of Earth where life exists.

CHAPTER → TWO

IS THERE LIFE ON OTHER PLANETS?

Scientists look for signs of life. They look for Earth-like planets around stars. They look for planets in habitable zones.

Mars is called the "red planet." Four billion years ago, it may have been habitable. It had rivers, lakes, and a deep ocean. It lost its **magnetic fields**. Magnetic fields are areas with moving electric charges. They protect planets from the Sun. The Sun stripped Mars of its life. Mars is now cold and dry. But scientists keep studying Mars.

Venus is dry. It's really hot. It has high air pressure. But billions of years ago, it had seas. Life on Venus hid in its clouds. These clouds have gases. These gases could give chemical energy to microbes.

Scientists found an exoplanet called GJ 357 d. This planet is 6 times more massive than Earth. It's 31 **light-years** away. A light-year is the distance that light travels in 1 year. Its air may be able to trap heat. This means it could make water on its surface.

Scientists think Earth rocks may have been blasted to Venus's surface.

K2-18b is another exoplanet. It's 110 light-years away. It has water **vapor**. Vapor is gas. But K2-18b doesn't have breathable air. It doesn't have a solid surface.

Teegarden's Star is a star. It is 12 light-years away from our solar system. Two planets orbit this star. Scientists believe these exoplanets could be habitable. Their air is similar to Earth's.

All planets, including exoplanets, form from clouds of gas and dust leftover from the birth of stars.

DOWN-TO-EARTH EXPERIMENT

Want to learn more about how scientists look for life in the universe? Try out this experiment! Think like a space scientist.

Materials:

- 3 clear cups
- Sand
- Sugar
- 1 package of active dry yeast
- Crushed fizzing tablet
- Mixing sticks
- Warm water

Instructions:

1. Fill each cup ¼ full of sand. Add a spoonful of sugar to each cup. Mix the contents.
2. Mark cups A, B, and C.
3. Do nothing with cup A. This should only have sand and sugar.
4. Add the crushed fizzing tablet to cup B.
5. Add yeast to cup C.
6. Mix each cup well. Use different mixing sticks.
7. Look for life in these 3 soil samples. Use your sense of touch, smell, and sight. Write down what you observe.
8. Fill up each cup halfway with warm water. See what happens. Water gives life. The tablet and yeast should move. Yeast is a fungus. It grows from the sugar. It gives off carbon dioxide, which makes foam. Foam and bubbles could be signs of life. Warmth is also a sign of life.

It's hard to find life. Rovers are sent to find life on other planets. Think about how to define life. What makes a living thing different from a nonliving thing? Scientists have a hard time defining it.

CHAPTER → THREE

IS THERE LIFE ON MOONS?

Billions of years ago, Earth and another planet crashed. Material flew. Earth's Moon was formed. Today, Earth's Moon is dry. It doesn't have air. It has dark **craters**. Craters are holes in the ground. These craters may have held ice water. The little water left is trapped in ice or rock. Traces of water may have come from underground. No trace of life has been found.

Jupiter has 79 moons. One of the moons is Europa. Europa's surface is mostly ice. The ice protects it from the Sun. It'd be hard to live on Europa's surface. Europa has no air. It's freezing. But it has a warm core. It has warm oceans running inside. Microbes could live in this warmer core. Europa may spit things out. This causes red spots on Europa. These red spots are rich in salts. They may be frozen bits of microbes.

Life on Europa would be like living in the freezing lakes of Antarctica.

Saturn is famous for its rings. It also has many moons. Titan is Saturn's largest moon. It's colder than Earth. This makes life unlikely. But scientists think Titan has chemical activities. Titan's air has gases. Titan is known to have water under its icy shell.

Enceladus is another one of Saturn's moons. It has an icy surface. It has an ocean underneath. Water shoots out into space. This water may have microbes that could contain life.

Asteroids, **meteorites**, and **comets** are space objects. They fly through space. They may crash into things. They make chunks of rocks. These rocks land on planets and moons. They may carry microbes.

The first astronomers thought the rings were moons. Astronomers are people who study space science.

IT'S (ALMOST) ROCKET SCIENCE

The Allen Telescope Array (ATA) is in the Hat Creek Observatory. It's in Lassen Peak, California. It's used to see things in space. It's used to search for extraterrestrial intelligence. It listens for signals from outer space. It has an array of 42 radio antennas. Arrays are groups. These antennas are very sensitive. They're reliable. They can hear areas around 20,000 red dwarf stars. Red dwarf stars are stars smaller than our Sun. They're the stars closest to Earth. As an array, ATA can do searches every day. It can search several different planets at the same time. It can be added onto. This speeds up the search process. It gets us closer to learning more about space. ATA wants to be the world's largest and fastest observing instrument. It's one of the world's most powerful telescopes. ATA's goal is to have 350 antennas.

CHAPTER → FOUR

WHAT IS TABBY'S STAR?

Tabby's Star is 1,470 light-years away. It's an odd star. It has strange dips in brightness. It glows brightly. Then, it **dims**. Dim means to get less bright. Tabby's Star kept doing this. It did this for several years. Scientists thought this was really weird. They couldn't figure out what was happening.

At first, they thought an exoplanet was blocking Tabby's Star's light. But dips happened during odd times. This meant the dips didn't fit a planet's orbit.

Another idea was a dust cloud. Tabby's Star wasn't new. So, dust clouds didn't make sense. Dust clouds only formed when new stars formed.

Tabby's Star is in the Cygnus constellation. Constellations are groups of stars that make a shape.

People also thought it was a comet shower. But this didn't make sense either. Thousands of comets would be needed to cause the dips.

Some people think Tabby's Star is part of an alien **megastructure**. Megastructures are large and complex arrangements. This alien group is thought to be smarter than us. It has advanced technology. Advanced means far ahead. The alien group draws power from Tabby's Star.

Tabby's Star is a mystery. It's named after Dr. Tabetha Boyajian. Boyajian is an American astronomer. In 2015, she found the star. She first noticed its strange light pattern.

Tabby's Star can only be seen with special telescopes.

CHAPTER → FIVE

HAVE ALIENS MADE CONTACT?

Some people believe aliens have made contact with us. For example, there are crop circles. These are strange patterns made in farm fields. They appear overnight. They appear mysteriously. People think aliens made them.

People believe there have been close **encounters**. Encounters are meetings. Close encounters have different types. There are close encounters of the first kind. This is the sighting of a **UFO**. UFO is an alien spaceship. There are close encounters of the second kind. This is a UFO sighting with proof. There are close encounters of the third kind. This happens when humans have contact with aliens. There are close encounters of the fourth kind. Examples are **abductions**. Abduct means kidnap.

UFO stands for unidentified flying object.

Betty and Barney Hill said aliens kidnapped them in 1961. They said the aliens had gray skin. They called them "grays." Some people think grays are at Area 51. Area 51 is a Nevada Air Force base. It's believed to have top-secret files.

Roswell is famous for its close encounters. It's known as ground zero for UFO research. Ground zero means starting point. Roswell is in New Mexico. In 1947, a UFO was thought to have crashed there. The government said it wasn't a UFO. But some people didn't believe them. They thought the government was hiding information.

People want to believe in aliens. But scientists say there's no proof of aliens.

Aliens have been used to explain many mysteries.

CHAPTER → SIX

WHO PROMOTED THE STUDY OF EXTRATERRESTRIAL LIFE?

Dr. Frank Drake was born in 1930. At age 8, he showed interest in life beyond Earth. He studied radio astronomy. Radio astronomy is a space science. It's the study of radio waves from space.

Drake used the Arecibo radio telescope in Puerto Rico. In 1974, he tried to contact aliens. He sent a digital message. He described Earth. He sent it to M13. M13 is a cluster of stars. It's 25,000 light-years away. Drake is still waiting for a response.

He created the Drake **Equation**. Equations are math statements. They explain things. The Drake Equation figures out the number of planets hosting extraterrestrial life. Drake guesses there are about 10,000 advanced alien groups.

Dr. Frank Drake studied astronomy at Harvard University.

The Drake Equation helps define extraterrestrial life. It considers the formation rate of stars. It considers habitable zones. It gets people thinking about life in space.

Drake helped develop the SETI Institute. SETI stands for Search for Extra-Terrestrial Intelligence. It started in 1984. It's in California. Scientists and teachers work there. They study life in the universe. They answer the following questions: What is life? How does it begin? Are we alone?

They think advanced extraterrestrial life knows how to use technology. These alien groups would use tools. SETI scientists build tools to search for alien life. They listen for radio signals.

NASA is also interested in searching for extraterrestrial life.

GLOSSARY

ABDUCTIONS (ab-DUHKT-shuhnz) kidnappings

ASTEROIDS (AS-tuh-roidz) small, rocky bodies orbiting the sun

ASTROBIOLOGY (as-troh-bye-AH-luh-jee) the study of life in the universe

COMETS (KAH-mits) space balls of ice and dust

CRATERS (KRAY-turz) dips or holes in the ground

DIMS (DIMZ) gets less bright

ENCOUNTERS (en-KOUN-turz) meetings

EQUATION (ih-KWAY-zhuhn) mathematical statement used to explain how things are equal

EXOPLANETS (EKS-oh-plan-its) planets orbiting stars outside of the Milky Way

EXTRATERRESTRIAL (ek-struh-tuh-RES-tree-uhl) being from someplace outside Earth

HABITABLE (HAB-ih-tuh-buhl) being able to support life

LIGHT-YEARS (LITE-yeerz) the distance that light travels in 1 year

MAGNETIC FIELDS (mag-NET-ik FEELDZ) areas around planets filled with moving electric charges

MEGASTRUCTURE (MEG-uh-struhk-chur) large and complex arrangement

METEORITES (MEE-tee-uh-rites) pieces of rock or metal that have fallen to the ground from outer space

MICROBES (MYE-krobez) tiny organisms

ROVERS (ROH-vurz) moving robots that explore the surface of space objects

SOLAR (SOH-lur) relating to the sun

THEORY (THEER-ee) an idea meant to explain something

UFO (YOO EF OH) unidentified flying object like an alien spaceship

VAPOR (VAY-pur) gas

FAR-OUT FACTS

- From 1952 to 1970, the U.S. Air Force studied UFO sightings. The study was known as Project Blue Book. It collected over 12,000 UFO sightings. Over 90 percent were "identified." This means they were caused by something known. About 6 percent were "unidentified." This means the cause was unknown.
- NASA sent baby squid to space. The squid traveled on the Endeavor in 2007. They have microbes that make them glow. NASA sent newly born squid. These squid didn't have the microbes yet. Microbes from the spaceship were put into the squid. They spent 28 hours in space. They were killed. They were brought back to Earth. Scientists studied them. They wanted to see the effects of space on microbes.
- In 2017, scientists saw an odd object. They thought it was traveling from another solar system. They called it "Oumuamua." This means "first messenger." The object was a few hundred feet long. It was thin. It was dull red. Some people thought it might be an actual alien spaceship.

LEARN MORE

Kenney, Karen Latchana. *Breakthroughs in the Search for Extraterrestrial Life.* Minneapolis, MN: Lerner Publications, 2019.

O'Brien, Cynthia. *Searching for Extraterrestrials.* New York, NY: Crabtree Publishing, 2019.

INDEX

aliens, space, 6, 22–25, 31
Allen Telescope Array (ATA), 17
antennas, 17
Area 51, 24
asteroids, 16
astrobiology, 7, 8, 9

"big bang," 4–5, 6
Boston, Penny, 9
Boyajian, Tabby, 20

close encounters, 22–24
comets, 16, 19
constellations, 19

Drake, Frank, 26–28

Earth, 4, 6, 8, 9, 10, 11, 14, 15, 17, 26
Enceladus, 16
Europa, 14, 15
exoplanets, 8, 11–12, 18
experiment, 13
extraterrestrial life, 6–9, 10–13

Jupiter, 14

life on other planets. *See* extraterrestrial life
light-years, 11

magnetic fields, 10
Mars, 9, 10
megastructures, alien, 20
meteorites, 16
microbes, 6, 10, 14, 16, 31
Milky Way, 4, 5, 8
moons, 14–17

planets, life on, 10–13

radio astronomy, 7, 26
Roswell, NM, 24

Saturn, 15–16
SETI Institute, 28
stars, 6, 12, 17, 18–21, 26, 27

Tabby's Star, 18–21
Teegarden's Star, 12
telescopes, 7, 17, 20
Titan, 15

UFOs, 22–24, 31
universe, 4–5

Venus, 10, 11

water, 6, 12, 14, 15, 16
women, 9

ABOUT THE AUTHOR

Dr. Virginia Loh-Hagan is an author, university professor, and former classroom teacher. She wrote the 45th Parallel Press books *Roswell, Zombies vs. Aliens,* and *Lost in Space Hacks*. She lives in San Diego, California, with her very tall husband and very naughty dogs.